Essential Reality

&

Time

Also by
Wonchull Park & Mackenzie Hawkins

Way of Now: Nowflow for Meditation,
Peak Performance, and Daily Life

Nowflow Breath, Movement & Mind:
A Living Practice of 3 Nowflow Qualities
from 3 Physics Flow Natures

Essential Reality

&

Time

Physics and Living of Nowflow

Wonchull Park
Mackenzie Hawkins

TH
RU

Publishing

Princeton, New Jersey

ISBN: 978-1-949706-07-9
Library of Congress Control Number: 2024910409

www.thrupublishing.com

Contents

Preface

If you've ever wondered about how *now* is so unique compared to past and future or wondered about the nature of time, this book is for you. You may not have a science background or you may be a physicist.

In history, some Eastern and Western philosophers have said that everything is now. In metaphysics, presentism represents this "everything-is-now" view, in contrast to the eternalism of 4-dimensional spacetime.

Presentism, however, faces objections because of some incompatibilities with physics. In this book, we introduce essential reality, which is based on the initial conditions of physics. We'll see how this essential reality, which is at a moment that includes the flow of changing, can resolve these objections to an "everything-is-now" view. From essential reality, we will also find a new interpretation of time, which is fully included in now.

We'll see the usefulness of essential reality in everyday life. For convenience in daily living, we can also call it nowflow, meaning now that includes the flow of changing. This book addresses some fundamental questions in physics and metaphysics, and applies them to daily life. We hope such applications are helpful not only to non-scientists but also to scientists connecting concepts with tangible experience.

Wonchull Park
Mackenzie Hawkins

Princeton, NJ
May 2024

Introduction

*T*his is a short book about essential reality. Instead of trying to figure out what is reality, which is a hard problem, we can introduce essential reality, which comes from physics and can be usefully applied in living.

Essential reality, as is explained in the chapters that follow, comes from physics since it is what's "just enough" to start with in physics (called initial conditions). Nothing more than essential reality is needed to account for effects in physics models and nothing less than essential reality would be sufficient.

One of the more striking aspects of essential reality is that only a single moment, and nothing more, is needed. This moment *now* is enough in this way. To be sufficient, though, the now must include the flow of changing at this moment. A moment as anything less than now that includes the flow of changing would be insufficient, as essential reality from physics makes clear. For convenient usage in daily living, we can combine the words *now* and *flow* to call it *nowflow*.

The applications for the essential reality of nowflow are quite widespread. Giving more attention to essential reality as we move can improve the efficiency of our movement, such as standing up quickly. We'll also see how trying to manage something non-essential can cause us to mismanage our actions. Though essential reality is based upon physics, we can apply it in our everyday living.

We can also apply essential reality to questions about reality, such as questions about presentism, eternalism, and the nature of time in metaphysics. The view of presentism is that reality is just at the present moment. This can be an intuitive view to live by. We cannot interact with the past that is gone or the future that is not here yet. Time beyond this moment is also beyond the reach of our direct experience. From the practical perspective of managing our lives, the view of presentism is simpler and more consistent with our experience of living.

Nevertheless, presentism faces some significant objections, perhaps especially from Einstein's relativity and 4-dimensional (4D) spacetime in physics. Could applying essential reality to this question of presentism resolve these objections?

Let's look again at a few points about essential reality that we've already touched on in this introduction. First, essential reality comes from initial conditions in physics. Second, since initial conditions are at one moment, essential reality is at one moment so it's like presentism in that way. Before going further into any details, these two points hint at how essential reality could potentially resolve apparent incompatibilities between physics and a presentism-like view where the present moment is all that's required.

In this book, we will go into how presentism-like essential reality is a valid option that agrees with physics because it has the full initial conditions, including the flow of changing. While this consistency is easier to see first in deterministic models, the applicability of presentism-like essential reality will later be generalized to indeterministic cases. We'll also see how essential reality doesn't require time, such as the dimensional time of 4D spacetime. Instead, there is an alternative explanation of time since all the effects of time come out of the changing that is included in essential reality. In seeing that this moment, which includes the flow of changing, is essential reality, we can find fuller meaning in how just this seemingly fleeting present moment is enough.

While some familiarity with basic physics would be helpful, it's is not necessary for understanding essential reality. The physics is described without formulas, even as all the physics is intended to be rigorous.

1

Essential for Whole Effect

What is reality is not an easy question. For example, is a past event part of reality? Even for what is present, we could ask whether velocity is reality—and what about acceleration?

We could also ask these kinds of questions not for all of reality, but for just a subset of reality instead. What if we could find some subset of reality which is essential because it clearly has its own effect? Perhaps this subset could be quite a small subset. Perhaps this subset could even account for everything else which has an effect.

That may sound hard to imagine. As we'll see, though, such a subset exists in physics models as initial conditions. Initial conditions are a small subset within physics models where each element of the set is *necessary* since it has its own effect. Also, once initial conditions are given in physics models, no additional effect can be added since the set is already *sufficient* for the whole effect. The subset that is necessary and sufficient for the whole effect in physics models is what we'll call essential reality.

There are various models in physics, such as the Newtonian model as well as models for relativity and quantum mechanics. These models all have initial conditions as a small, essential subset that is necessary and sufficient for the whole effect. (Indeterministic interpretations of quantum mechanics will be considered later.) Examples of initial conditions in simple Newtonian physics models are given below. These examples also show how the whole effect covers the effects of moments in the past and the future when that moment was or will be now.

Examples of initial conditions

As a very simple example, let's look at the initial conditions in a Newtonian physics model describing a world with just one moving ball. Without any interaction, this one ball moves forever at the same speed in the same direction (i.e., the same velocity). From this, we can see how the small set of position and velocity at a moment is sufficient to cover the effects at all moments (i.e., the whole effect), including the ball's past and future trajectory.

As a next example, we could consider a world
with two balls interacting through the gravitational attraction of their mass. What would be
the initial conditions in the Newtonian physics
model describing this world? The position and
velocity of things (with mass) at a moment are
still the initial conditions that cover the whole
effect, including the balls' past and future.

We can connect this to our experience in a
case when one of the two balls is much, much
larger. When we throw a ball, for example,
we're throwing a smaller ball on the much,
much larger ball of the Earth, which has its
mass or can be equivalently accounted for as
Earth's gravitational force. (Using a constant
gravitational force on the ball means Earth's
initial conditions enter through this force,
and subsequent movement of the Earth is
negligible so this force remains constant.)
From such experiences, we know that the
future of where the small ball goes depends
on how the ball initially leaves our hand. If
those initial conditions are different, such as
having a different angle or speed (i.e., different
velocity), the ball will have a different trajectory. Each initial condition contributes its own

effect which is needed for the whole effect.
Also, adding something else, like the ball's
acceleration at the moment the ball leaves the
hand, doesn't contribute its own effect on top
of what is already given by the essential set of
initial conditions. [Ref 1]

At a moment & changing

These examples of initial conditions illustrate a
couple of points about essential reality. First, just
one moment can be essential reality because initial
conditions are just at one moment. In the simple
examples we looked at, the initial conditions are the
position and velocity of things (with mass) at only
one moment. Just one moment that includes the full
initial conditions is sufficient to cover the effects
at all moments (i.e., the whole effect) in physics
models.

Since a moment is so brief that may be rather
striking. Essential reality is not spread out in time
across two different moments, such as including
events from a past moment as well the present
moment. As initial conditions make clear, past
events are not needed to cover the whole effect that
is already accounted for with initial conditions at

just one moment. We can choose the one moment of essential reality to be the present moment since our experiences of reality are centered in the moment that's now.

A second point about essential reality is that it includes the changing, or velocity, at a moment as well as the position of things. Now as "now-shape" with just the static position, or shape, of things at this moment would be *insufficient* for the whole effect since it's missing the initial condition of velocity. Even in the simple examples we've seen, the velocity of the ball at an initial moment must be included for the whole effect to be covered.

Now is *sufficient*, though, when it also includes the flow of change at this moment. To indicate this, we can combine the words "now" and "flow" into one word, nowflow, which can be useful when applying to our living. The essential reality of nowflow has what is necessary and sufficient for the whole effect in physics, and we'll see how to apply this in living.

Just enough
As we've seen, the essential reality of nowflow is based upon physics models. It's coming from initial

conditions and how they are the minimal subset
sufficient for the whole effect. When physicists and
engineers use initial conditions, they deal with this
small yet sufficient subset to describe and manage
actual things and situations.

It's quite useful to be able to manage the variety and
complexity of actual things and situations with a
manageable, small subset. Essential reality allows us
to see what's just enough (i.e., necessary and sufficient for whole effect) according to physics. It shows
us viable options for simpler views. By seeing this
in physics, we can then apply the essential reality of
nowflow in our living, such as our physical movement.

2

Effects in Action

T here can be many effects to manage in a physical movement. When throwing a ball, for example, there's the trajectory of that ball's position over the past, present, and future. There's also the ball's velocity and the ball's acceleration. Here we'll apply the essential reality of nowflow to physical movement and see how taking care of essential reality is what we need to manage physically.

As we saw in the last chapter, the position and velocity of things (or accounted for as the equivalent force of interactions according to that) are part of initial conditions. In this way, position, velocity, and force are included in essential reality. By contrast, acceleration is not an initial condition and so doesn't add an additional effect not already covered by essential reality. In this chapter, we'll contrast acceleration, which is non-essential, with essential reality in our experience of movement. This will also help us connect the concept of essential reality to our tangible experience.

Force & acceleration

Let's say we're a passenger in a car that's getting up to speed after a stop sign. Through the window, we could see that we were moving slower a moment ago and are presently moving faster, which is one way to know that we're accelerating. (This is basically the definition of acceleration, which is change of velocity in time.)

We could also think that we are feeling the car's acceleration more directly at a moment. As the car speeds up, we're feeling the seat pushing our back in the moment and may be used to associating that with the car's acceleration. Looking into this experience, though, we see that the pushing into our back is actually a force that we're feeling, and the acceleration is inferred from that force as we'll see more clearly in the next example.

Let's say that we're whirling a ball on a string in a circle over our head by pulling on the string with a force from our hand. Though we'd feel the pull of that force at our hand as we whirl the ball, we might not think about how we're accelerating the ball in this situation.

The acceleration is less obvious in this case because the direction of movement is changing (rather than the speed going faster or slower). We may need to recall some high school physics to know that the ball is accelerating inward towards the center of the circle for it to do this circular motion. It's not necessary, though, for us to be aware of the ball's acceleration for our whirling action as long as we're managing the inward pulling force of the string in our hand.

This is an example of how we can manage our movement through feeling force since, by managing the force, the acceleration is also taken care of. This is consistent with acceleration as non-essential and so it doesn't add an additional effect. We would likely mismanage our movement, though, if we're thinking about acceleration and not feeling the forces now in our movement, as we'll see in the next example.

Managing (and mismanaging) nowflow

Let's say that we need to stand up very quickly. Perhaps a sudden noise has startled us and we're getting up to deal with whatever might be the problem. In a situation like this, we might strain our head and upper body upward to try to accelerate upward faster.

In these kinds of situations, we sometimes react with an overly strong intention to accelerate fast. The fact that we're straining upward with our head and upper body is a clue that we've disconnected from feeling the force of the ground at our feet and so are missing something important for this action.

It's the force now, especially the force between our body and the ground, that gives our upward acceleration. If we were feeling the forces in this action, we'd likely lean forward first (rather than strain upward) so that more of our weight is over our feet faster, allowing us to push off the ground more effectively. We can experiment with this for ourselves by standing up in two ways.

In one way, we could quickly get up from a chair thinking mostly about accelerating upward faster. When we are over-focused on acceleration and don't manage the forces at work in our action well, our acceleration would be less likely to happen effectively. In comparison, we could next quickly get up from a chair by leaning forward and feeling the force of our feet pushing off the ground. By reminding ourselves to feel the force, the accelera-tion upward as we stand would still be taken care of and likely done more quickly. We could practice standing up, or other movements, with more feeling of the force of things interacting now.

Force flowing form

As a reminder to feel and manage the essential reality of nowflow in our movement, we can use the phrase *force flowing form*. With *flowing form*, we remind ourselves how our body isn't a "now-shape" form. As indicated by initial conditions, both our body's position and velocity at this moment are part of the essential reality of nowflow.

As we stand, the force of interaction is already part of the *flowing forms* of our body and the Earth. Interaction with the Earth, though, can also be usefully represented as a force, such as the force exerted by the Earth on our body, so the reminder of *force flowing form* is practical in this way.

By directing our attention towards managing the essential reality in our movement, the reminder of *force flowing form* can help us manage our movement of standing. When we lean forward so that more of our weight is over our feet faster, it's the *flowing form* of our body that is increasing the *force* at our feet. That *force* of our weight then helps to change the *flowing form* of our body, accelerating it upward. By feeling and adjusting nowflow's *force flowing form*, we are taking care of our action, including our body's acceleration.

We could also apply this to other situations of movement, such as running. Perhaps we're in a race and are just about to be passed by another runner. Instead of an overfocus on accelerating (and possibly straining ineffectually), runners could direct their attention more to the source of running faster. Each time their foot hits the ground, the *force flowing form* now is what's needed and sufficient for the acceleration of their body.

As we've seen in these examples, taking care of nowflow's *force flowing form* is enough for the effect of acceleration in our physical action. By applying the essential reality of nowflow to physical movement, we could see and manage our movement more simply in a way that is consistent with our own experience of movement as well as with physics. Next, we'll take account of mental experience, such as when we're dealing with past, present, and future in managing our lives, and see again what is essential for us to take care of.

3

Past, Present, Future

*I*n the last chapter, we saw examples of how nowflow's *force flowing form* is enough for the physical effects of movement, including acceleration. For the movement of our body, it might be quite evident in our experience that this moment, now, is all that we can and need to manage physically. In our lives, though, it may seem like there's more than just this moment now that we need to deal with.

The events of the past and the future are outside of this present moment, so, as we go about our lives based on what has happened in the past and preparing for the future, it may seem like we're needing to deal with something outside of the essential reality of nowflow. In this chapter, we'll look more fully at our experience, including our mental experience, to see what is really enough for us to manage in an everyday example of walking to a destination.

Story past, present, future

As we're walking to a destination, our physical action is nowflow's *force flowing form* with each step now. Step by step, our movement is always this *force flowing form now.* As we walk, we also have a destination in mind that is guiding our steps now. There's a plan with information in our mind now about where we want to be in the future.

This plan for our destination is a story about the future so we can call it the *story future.* The story future about our arrival at the destination is happening in our mind now, in contrast to the future event itself of our arrival, which hasn't happened yet.

As we walk along, the future event itself of our arrival *isn't now,* since it's in the future. The story future, though, *is now* as information about the future that is guiding our steps now. A similar distinction can be made for the past in terms of the past itself and the *story past.*

As we walk along now, the past event itself of our departure is gone from our experience now since we are no longer at the place we started anymore. That past event itself is gone and is no longer in our experience. What is part of our experience now is the memory of our departure as a story past happening now in our mind.

So far, then, we've found in our experience that there is information about the past and the future happening now in our mind as a story past about our departure and a story future about our arrival. We also have a story present about where we currently are. Together, this story past, story present, and story future is information about our walk so that we know what has happened, what is happening, and what will happen. This *story past, present, and future* can be part of our experience now as we walk along because it's information in our mind now.

What we've not found is something outside of this moment, now, that we must deal with as we walk to the destination. What is outside of now isn't even available for us to experience or interact with. For example, we won't actually deal with the future event itself of our arrival until we arrive there and that event becomes now. Until then, it's the story future (happening now in our mind) about our arrival that we need and can deal with.

Though the past and the future in our experience may *seem as though* there's something outside of now that we're experiencing, what we find is that this actually isn't so. This is consistent with how the past and the future are non-essential because past and future events are outside the essential reality of nowflow. We will get to where we will be in the future by managing just what is now, which includes the story future for our arrival. Let's also see how we'd likely mismanage our experience if we feel more disconnected from the essential reality of nowflow.

Managing (and mismanaging) nowflow

Let's say that we're hurrying along a street to get to an important meeting and we're running late. In trying to get to our meeting on time in the future, we might be less aware of our present situation. We might even step into a busy street without seeing an oncoming car. This is an example of how we may miss something important for getting to the meeting, such as the present perception of an oncoming car, if we're feeling more disconnected from nowflow.

As this example shows, we may sometimes feel *as though* the future itself, like the future event of our late arrival, partially separates us from what's happening now. This sense of partial separation from nowflow makes it harder to manage what's happening now, even though the story future about our arrival is actually happening now in our mind along with the story present of the oncoming car. Since this sense of separation from nowflow only seems to be happening (but isn't actually), we could call this the illusion of separation from nowflow.

We can also have the illusion of separation from nowflow if we're trying to stand quickly, as in the example from the previous chapter. Standing up quickly is like a shorter version of hurrying to a destination. Even for this short movement, our movement is being guided by a plan, or intention, in our mind now for our future position. In hurrying to get to the standing position in the future, we may strain upward with our head and neck, instead of feeling and managing the force now at our feet, for example. This indicates how we're feeling partially separate from nowflow's *force flowing form*, which is what we actually need to take care of to stand.

Lessening the illusion of separation from nowflow would make it easier for us to manage what really matters in standing up, which is the essential reality of nowflow, so that we could stand up more quickly and effectively. This applies to getting to the meeting as well. With less illusion of separation from nowflow, we could more speedily and safely maneuver through the pedestrian and street traffic happening now. We could also choose the best routes by using our story past, present, and future (in our mind now) to get to the meeting on time if possible.

Living in the moment

When we look at our experience and find the story past and the story future, we see how we aren't really dealing with the past itself and the future itself in our experience, even if it might seem so at times. What we can and need to manage in our living is just what's going on now. This is consistent with how past events and future events are outside of the essential reality of nowflow and are non-essential since they don't add an additional effect.

That we can only live in this present moment is, in a way, common sense from our experience. Advice encouraging us to "live in the moment" and "be here now" also makes sense since we can sometimes feel as though we're partially disconnected from the present moment and that can cause us difficulty. As we saw in the examples of hurrying to a meeting or standing up quickly, the illusion of separation from nowflow can make us miss something important for our actions.

"Being now," though, is not something that we have to try to achieve since we never actually disconnect or separate from nowflow. When we are being affected or affecting something, that takes place in the now. Even the story past, present, and future as information in our mind is part of what is happening now. Some people in history have said, "Everything is now." Further explanation and various applications are discussed in the book *Way of Now: Nowflow for Meditation, Peak Performance, and Daily Life*.

As we'll see in the next chapter, there is long-standing debate in metaphysics over whether just the present moment is enough to be all of reality or whether reality needs to include more than only the present moment. We'll next apply essential reality to these questions in metaphysics about the reality of the past, present, future, and time.

4

Presentism & Eternalism

*I*n this chapter, we'll look at how we can apply essential reality to questions about what is reality and what isn't. In metaphysics, there are two main views regarding the reality of the past, present, and future. One view, called presentism, is where only the present moment is reality. The other main view is eternalism, such as seeing reality as an eternal block of 4D spacetime that includes the past and future. [Ref 2]

The view of eternalism and 4D spacetime is seen as more consistent with physics than presentism, which faces objections regarding its compatibility with physics. Presentism, though, is more consistent in some respects with our experience. As we saw in the previous chapter, we experience events themselves only as present events and can't interact with the past or future itself, so the view of presentism matches our experience more simply.

The essential reality of nowflow is also only at the present moment so it's like presentism in that way. Essential reality, though, includes more than presentism since it's the full initial conditions at a moment, including velocity at that moment. As we've seen, essential reality also covers the whole effect in physics models, so essential reality is also consistent with physics and views of 4D spacetime.

This consistency can also be seen through an equivalence-principle-like reasoning. As long as essential reality is included in any views of reality, physics cannot distinguish between these views of reality. So they are equivalently valid views of reality according to physics. It is a kind of equivalence principle, but about reality views, so could be called reality view equivalence.

In applying essential reality to questions about the reality of the past, present, and future in metaphysics, we can consider the view where we take reality to be only essential reality. In this view, reality is only at the present moment like presentism, yet it's based on essential reality, so we can call this view "presentism-like essential reality." Below we'll see objections to presentism and how presentism-like essential reality answers these objections.

Objections to Presentism

Here we'll look at three main objections to presentism. These objections relate to truth, to relativity, and to time. For each, we'll also look at presentism-like essential reality and whether it can provide resolution for the objection.

Truth & the past

One objection to presentism is that, if only the present moment is reality, there seems to be issues with knowing the truth of what happened in the past. At the present moment, there can be information *about* what happened in the past or evidence *for* what happened in the past. Even so, there doesn't seem to be enough inside reality, if it's only at the present, to know whether some past is true or not (called truthmaker). [Ref 3] In applying essential reality to this objection, we can see how it matters whether or not the present moment includes the initial conditions.

A "now-shape" moment is missing the initial condition of velocity, so reality as a now-shape moment wouldn't be sufficient to know whether some past is true or not in physics. If we take the present moment to be the essential reality of nowflow, though, this is the full initial conditions, including velocity at a moment. As we've seen, the essential reality of nowflow is sufficient to cover the whole effect in physics models, including the past and future. With presentism-like essential reality, what is at the present moment is sufficient to know the truth of what happened in the past in physics models, which in principle provides a solution for the present being sufficient to cover the truth of what happened in the past.

(Note: Quantum mechanical models are included in this reasoning. Wave functions can be used in this reasoning with an interpretation that supports their deterministic meaning. However, interpretations of quantum mechanics are far from uniform. If one prefers an indeterministic interpretation of quantum mechanics, please see the final section of Ch. 5.)

Relativity
Another objection to presentism is that the present in Einstein's relativity depends on the observer. One observer might see some events as happening simultaneously at the present and another observer (moving at a different velocity) might see the same events as not happening simultaneously. [Ref 4, 5] As an

example, one observer's present event can be another observer's past event. This may seem to necessitate something more than the present of an observer to cover all of reality.

The essential reality as defined before (i.e., the necessary and sufficient conditions for the whole effect) also applies to relativity and can clarify this situation. Presentism-like essential reality has the full initial conditions so that, once the essential reality of one observer is given, this covers the whole effect of 4D spacetime. This further covers every other observer's essential reality and 4D spacetime.

Returning to the example when a present event for one observer is a past event for another observer, this is an aspect of presentism-like essential reality being dependent on the observer. However, these two observers' essential realities mutually determine each other, so there is no addition needed beyond the essential reality of one observer to account for the other observer's essential reality.

This can be more clearly seen by returning to the reality view equivalence. For an observer, once its essential reality is given, whether another observer's essential reality is added as additional reality or not cannot be distinguished by relativity physics. This equivalence shows that an observer's essential reality remains the necessary and sufficient condition for the whole effect.

What is common to all observers is a 3D essential reality unfolding in 4D spacetime. Since the essential reality of any one observer covers all other observers' essential reality and how it unfolds in 4D spacetime, it is consistent to state that the presentism-like essential reality of an observer is the observation of the whole reality for that observer. Since one observer's essential reality also determines any other observer's observation, that a different observer observes differently does not change this fact of consistency.

In this way, any observer has its presentism-like essential reality that covers every observer's essential reality and 4D spacetime. This can be interpreted as the observation of reality depends on the observer. As an example, two events can change order depending on the observer. Such change in order can only happen, though, for events without possible causal connections (i.e., where even light cannot connect the two events). Since causality is preserved in relativity, this indicates an underlying consistency of various observer's observations, including observations of presentism-like essential reality. This can be expected because an observer's essential reality determines any other observer's essential reality.

In summary, for an observer, experiment cannot distinguish between the following reality views: only its own essential reality, another that adds the essential reality of other observers, and another that further adds the entire 4D spacetime. (In other words, each observer has its own essential reality. Since any one observer's essential reality determines all other observers' essential reality, an observer's essential reality can be considered the observation of whole reality.) So, it is consistent to consider the essential reality of an observer as the whole reality for that observer, covering the whole 4D spacetime effect. Therefore, presentism-like essential reality is valid in relativity as well.

Time

One of the long-standing questions about
presentism is that, if reality is only the present
moment, then what about the reality of time?
There also seems to be a self-contradiction in the
presentism view since it categorizes time into
past, present, and future and then would deny
that there is time, or any reality, beyond the
present moment.

We've already seen how the initial conditions at
one moment are sufficient for the whole effect
of 4D spacetime in physics models. However,
initial conditions include velocity which is
usually considered as a time derivative. Instead,
to start without time, the initial condition of
velocity can be considered as a ratio of *comparative changing*.

We start with comparative changing between
the changing things in essential reality, such
as one thing moving twice as fast as another
thing. That's a comparative changing ratio of 2,
a dimensionless number. For example, when a
merry-go-round is turning, a toy horse on the
edge of the rotating platform is moving twice as
fast horizontally compared to a toy horse that's
halfway to the center. (When velocity is used in

a scalar sense like it is here, it can be considered as each of the 3 components of the vector.)

For the current purposes, starting with essential reality as position and comparative changing and showing one way to recover the full time effect will suffice, while a more general explanation of time will be given in the next chapter. Below is a specific example of recovering the numerically same values for time starting from essential reality that includes comparative changing (instead of velocity as a time derivative).

Among the comparatively changing things in essential reality, we can choose one changing thing as a standard (i.e., a clock) to measure ratios of comparative changing. As an ideal example, let's choose a light clock, such as light bouncing between two mirrors. For this light clock, we can then choose a suitable standard unit so that the numerical values for its unit are the same as a usual time unit of a second. This works when we choose the unit for this clock to be its light traveling a distance of c meters (where c denotes the constant value that is the speed of light in meters per second). With the unit for this clock as light traveling c meters, the clock gives the same numerical values as the unit of a second in time.

With the numerically identical time values and ratios of comparative changing, there are also the same values for velocity as a time derivative. It then follows that initial conditions and the whole effect in 4D spacetime are also equivalent, so the effects of time are also covered. This demonstrates how starting first with essential reality that includes comparative changing is sufficient for the effects of time, instead of first depending on time derivatives.

Since the whole effect is equivalent in either view, it is the order of interpretation that is different whether one starts with a 4D space-time view or an essential reality view. Though we've seen this here through one example of choosing a particularly suitable clock and clock-change unit, we will look at this topic of time more broadly in the next chapter.

Essential & non-essential

Presentism-like essential reality, as we've seen, is like presentism since it's a view where reality is only at this present moment. Presentism-like essential reality differs from presentism, though, because the essential reality of nowflow also covers the whole effect of 4D spacetime in physics. In terms of covering the whole effect in physics, reality as essential reality is sufficient. As we've seen, this shows how presentism-like essential reality is compatible with physics.

In physics models, once essential reality is given, all the effects remain the same whether one includes the past and the future as part of reality or not. In this way, the past and the future are non-essential since the essential reality of nowflow is already sufficient for the whole effect in physics models. In the eternalism view, past events themselves and future events themselves are considered as part of reality in addition to essential reality at the present moment. In this view, reality is a much, much larger set than essential reality, although the whole effect is the same as presentism-like essential reality. The clarity and simplicity of presentism-like essential reality, which is a small set only at a moment, can be useful in various situations.

Another way of looking at this same reasoning is through the reality view equivalence. As long as a view of reality includes essential reality, physics cannot distinguish between those views, so they are equivalently valid. One example of this reality view equivalence is the equivalence between presentism-like essential reality and eternalism.

5

Time

*I*n this chapter, we'll see a broader context for considering essential reality and time. We'll see how the order of interpretation is different between starting with 4D spacetime, which has dimensional time, or starting with essential reality, which has comparative changing.

In addition to looking again at the specific example of the light clock, we'll look at more general examples for comparative changing and time values, as well as examples for space and length values. This will help us contrast starting with essential reality as position in space and comparative changing to the usual way of starting with 4D spacetime.

Space and time

For example, if we're deciding whether a sofa will fit in a room, it may be cumbersome to have to directly compare the spatial extent of the sofa and that of the room. By using a standard measure of spatial extent (i.e., a ruler), we can quantify the spatial extent of both the sofa and the room and compare these length values to see if the sofa fits. We're using length values from rulers to quantify and manage spatial extent.

Time values are similarly useful. Let's say that we're heading to a meeting and we want to know whether we'll be on time. By using the standard measure of a clock, we can see if the time duration needed to arrive at the meeting fits within the time remaining before the meeting begins.

As this example illustrates, we're often thinking about things happening in space and also in time, so we're constructing in our minds something similar to 4D spacetime. Though 4D spacetime might usually be associated with Einstein's relativity, other physics models also use a time axis that's a dimension of time, which we refer to as the dimensional time of 4D spacetime. Even in daily life when we think about whether we have enough time to get to the meeting on time in the future, we may be similarly thinking about dimensional time and whether there is enough of it before that future event happens.

In this way, our default view in daily living as well as physics may be that there is dimensional time and that time values from clocks are a measure of this dimensional time. This view starts with 4D spacetime as a given, which is why the chart below starts with space and dimensional time in the top row. In the left-hand column, the chart summarizes how space is quantified with a ruler to obtain length values. In the right-hand column, we see how this 4D spacetime view starts with dimensional time that is quantified with a clock to obtain time values.

4D Spacetime view
space | **dimensional time**
ruler | clock
length | time values

Since dimensional time, alongside space, is what we start with in this view, it may seem that dimensional time is indispensable. Clocks and time values would seem to be dependent on the reality of dimensional time. Velocity as well in this view is dependent on time as a time derivative, i.e., changing happening in dimensional time. Since velocity is an initial condition, it might even seem as though the dimensional time is required as part of essential reality.

In the last chapter, though, we already saw an exception to this view that dimensional time is indispensable. With the light clock, we saw an example of how essential reality (with comparative changing and without need of time derivatives) is enough for a suitable clock that gives identical time values and initial conditions as usual. Here we'll continue to look at the broader context for how essential reality is sufficient for the effects of time without needing dimensional time.

Starting from essential reality

Let's return to the example of the sofa and the room to see how space, rulers, and length values are part of essential reality. Essential reality, as we've seen, includes things with position in space. We can compare their spatial extent directly, such as bringing the sofa into the room to see if it fits. We can also choose one thing with spatial extent as a standard measure, i.e., a ruler like a meter stick, for quantifying every other spatial extent. With a ruler, we can measure the length of the room and the length of the sofa and compare the length values.

Similar to how a ruler is something in essential reality chosen as a standard measure, a clock is also something in essential reality chosen as a standard measure. Of the many changing things that are part of essential reality, we can choose a changing thing as a clock to serve as a standard measure for quantifying any other change. More precise clocks can be a light clock or a cesium atomic clock, for example. A certain amount of change in a clock becomes the standard unit, which is used for time values to quantify other changes.

In the summary below, the essential reality view starts with space and comparative changing in the top row of the chart. This is contrasted with the 4D spacetime view that we looked at previously which starts with space and dimensional time.

4D Spacetime view	Essential Reality view
space | **dimen. time**	space | **comparative changing**
ruler | clock	ruler | clock
length | time values	length | time values

In looking at this summary, we see how both views start with space where a chosen standard spatial extent can serve as a ruler to give length values. According to both views as well, a clock is a chosen changing thing that gives the standard measure of

time values. These two views are different in that
the 4D spacetime view starts with space and *dimensional time* whereas the essential reality view starts
with space and *comparative changing*.

One clock, two interpretations

With this summary of the two views in mind, let's
return to the example of the light clock introduced
in the previous chapter. We've seen how this clock
in either view can give the numerically identical
time values, so that there are the identical initial
conditions and the same whole effect in physics
models. We could say that's a quantitative way of
seeing the validity of both views. Since the two
views of the light clock can be numerically identical
and equivalent for all effects in physics, it is the
interpretation of the clock that differs between these
two views.

In the 4D spacetime view, it's a given that there is
dimensional time as part of reality. We then quantify
the duration of dimensional time by using a clock,
such as the light clock, that provides a standard
measure, such as a second of dimensional time.

As we've seen, we can also view the light clock with
a different interpretation. In the view of essential

reality, we start with what is necessary and sufficient for the whole effect. This means starting with the essential reality of space and comparative changing (without dimensional time). We then choose a comparatively changing thing, such as a light clock, as a standard measure for change that provides time values, such as a second, as information for comparing changes. In this view of the light clock, dimensional time is not required, even though all the effects of dimensional time in 4D spacetime are covered by essential reality, as we've seen.

There can be situations where the view of essential reality has advantages since it's simpler even though it covers the same whole effect as 4D spacetime. If one were to study a new effect of time, for example, it would be clearer in this view that one would study the behavior of a clock (as Einstein did by replacing time with clock in developing relativity) since time effects are fully given by the behavior of clocks. In the 4D spacetime view, dimensional time is assumed before clocks, so this is less obvious.

In our living, it is usually more effective to adopt the simpler view of essential reality. In the example of going to the meeting, we've seen how we can't actually interact with a future event or the dimensional time of the future, yet we may experience the illusion of separation from essential reality and so miss something that we actually can and need to manage, like oncoming traffic. Let's see what it could be like to manage getting to the meeting on time without a sense of there being dimensional time.

Managing essential reality

Let's say that we're halfway to our destination when we look at our watch and see the time value for the present. We also looked at our watch when we left home and so have a memory of that previous time value in our mind now. If we expect to continue at the same pace, we can also have a plan for what the time value will be when we arrive at the meeting.

Our story past, present, future and its related time values are helpful information available to us as we head to the meeting. We can see whether the time value in our plan for our arrival is less than the time value of when the meeting will begin. If it's not, then we'll need to hurry up so that we're not late, which is how time values help us manage change. This is similar, as we've seen, to how length values from rulers help us manage spatial extent, like seeing that the length values for the sofa are less than those for the room.

In the example of getting to a meeting on time, there are two changes that we're wanting to manage in relation to each other: The change of our position from where we are to the destination of the meeting place, and the change of the meeting organizers being ready to start the meeting. By quantifying both changes with a standard measure of change (i.e., a clock), these time values for the story future help us manage these changes.

In this view, time values can quantify the changes that have happened in the past or that will happen in the future. Those changes themselves as events are outside of essential reality since essential reality only includes chang*ing* at a moment (and not changes beyond that). Essential reality, though, includes available information about those changes regarding the past and future, including time values as information coming from clocks in essential reality.

As we've seen, we can quantify time duration by looking at a clock (when it's a changing thing in essential reality) as we depart and by using that information together with reading the clock again now as we are halfway to the meeting. Time values can quantify time duration as information that's included in essential reality. The time duration itself as change itself, though, includes the past event of our departure and so is outside of essential reality. By contrast, when we measure the extent of a sofa, we are quantifying something in essential reality. Spatial extent is part of essential reality in contrast to time duration itself, which is non-essential.

In getting to the meeting, we've seen how we need to manage the changing of this moment and our story past, present, and future about changes. For this, we can use a clock as chosen changing thing in essential reality now that allows us to have time values quantifying our story past, present, and future about changes. In this view, time is the standard change information (in essential reality) coming from a clock, which is the chosen standard changing thing (in essential reality).

This view of time, which we could call *essential reality time*, is simpler, since it doesn't span the past, the present, and the future like dimensional time. All essential reality time needs is space and comparative changing, and, as we've seen, that is also sufficient for the whole effect of dimensional time.

Essential reality time

Essential reality time, as we've seen, is fully part of the essential reality of nowflow, even as all the effects of dimensional time are covered. In this way, a view of time as essential reality time, which is simply the standard change information in essential reality, allows full use of time in our lives without the tendency for time to make us feel the illusion of separation from nowflow.

Having a sense of essential reality time as part of nowflow, for example, could lessen the illusion of separation as we hurry to get on time to a meeting. By feeling how all that we can manage is part of the essential reality of nowflow, including time, we could more easily manage the story future with its important time values and also take care of the story present with the oncoming traffic, since all of that is more clearly happening together now.

Over two thousand years ago, the philosopher Aristotle gave this definition of time: "It is clear, then, that time is number of movement in respect of the before and after." [Ref 6] Essential reality time includes several aspects of this definition. Numerical time values come from the movement, or change, of a clock. With time values, any other movement or change can be quantified in the story past, present, and future or in respect to the before or after.

As a definition of time, though, there is a further or differing aspect of essential reality time, which is that there is only moving and changing, rather than finite movement and changes across more than a moment. Since essential reality is only what it is, which is the now that includes the flow of changing, there is no before or after as actual events in essential reality.

In short, for presentism-like essential reality, comparatively changing things in space at the present moment is all of reality. A standard changing thing as a clock gives the standard change information as time to compare other changes. This has been shown to be consistent with physics. The presentism-like essential reality can be interchangeably called the nowflow view, reminding us in our living that now with the flow of changing (as well as the position of now-shape) is the whole reality.

(Note on velocity as intrinsic to a moment: The velocity as comparative changing can be naturally defined as an integrand, v*dt=dx, instead of the customary definition as a derivative, v=dx/dt. Although these give the same numerical value, the derivative definition requires velocity arising from more than one moment, while the integrand definition can be intrinsically at one moment. To solve physics equations, this velocity is integrated in time (change of clock) to give position, then position and velocity determine the force, and force is integrated to give velocity at the next moment. An intuitive example of comparative changing as intrinsic to the character of a moment can be seen with a rotating disc where the geometric structure of a point at the outer rim and at half the radius gives a ratio of 2 for the comparative changing.)

Generalization beyond deterministic cases

In coming to this nowflow view of reality, or presentism-like essential reality, deterministic cases were used for their clarity. In those cases, as we've seen, the whole effect (including all the effects of past, present, and future in 4D spacetime) is equivalently covered by the essential reality of nowflow. Having arrived at this nowflow view of reality, we can generalize the applicability of this view to indeterministic cases.

Let's consider an indeterministic case where the
initial conditions of position and velocity (or
position and velocity information for quantum
wave functions) still exist, but the initial conditions
don't completely determine a future event. That
these initial conditions exist means that they cover
all information available at the moment of initial
conditions about that moment and about the future.
In other words, no additional information can be
added to that moment of initial conditions. Even
in this indeterministic case, the initial conditions
of position and velocity are what's necessary and
sufficient (i.e., essential) *for that moment of initial
conditions*, which is the nowflow of that moment.
We can next apply this to whether the future event
would need to be added to essential reality.

It may seem that a future event not determined
by the initial conditions of position and velocity
would need to be added to essential reality, since
such a future event may seem to add its own additional effect not covered by the essential reality of
nowflow. However, when the future event has its
effect, it's not the future anymore but is now. As
we've seen, the initial conditions of position and
velocity are still what's necessary and sufficient (i.e.,
essential) *for that moment of initial conditions*. Whatever is at that moment is already covered by the

essential reality *for that moment*. Because the initial
conditions are position and velocity in this case, the
nowflow view still applies.

(In other words, the reality view equivalence still
applies even in this case when the future is indeterministic. The reality view equivalence states that,
as long as essential reality is included in any views
of reality, physics cannot distinguish between these
views of reality. A distinguishing effect such as in
an experiment can happen only in the now, and
the essential reality of nowflow is still necessary
and sufficient for each moment of the experiment.
For example, measurement processes that are the
same in essential reality will be equivalent, meaning
they do not introduce any different effect. For a
probabilistic physics model, the probablilities will
remain the same, even when individual results are
indeterministic.)

Similarly, when the essential reality of nowflow for
the present moment does not contain the whole
information of the past, this does not necessitate
an expanded essential reality. When the past event
had an effect, it was now. The initial conditions
of position and velocity at that moment are still
what's necessary and sufficient for that moment.
For example, the present may not contain the whole

information of a past event, but the essential reality of nowflow for that moment covers the event itself.

Both for a past moment and a future moment in an indeterministic case where the initial conditions are position and velocity, the nowflow view still applies since such a moment is covered by the essential reality *for that moment*. In generalizing the applicability of presentism-like essential reality, or the nowflow view, to indeterministic case, there still is the same view of reality as stated before: Comparatively changing things in space now is all of reality. A standard changing thing as a clock gives the standard change information (which is inside nowflow) as time to compare other changes.

6

Managing Essential Reality of Nowflow

*I*n our lives, we can't go back and interact with a past event, nor can we interact with a future event that hasn't happened yet. Our actions, as we manage our lives physically and mentally, always happen in the present moment. In this way, the view that "everything is now" would support the managing of our lives. It's a difficult question, though, to know what is the "everything" of reality. Is only one moment as all of reality enough to account for everything?

Here we've asked an easier question by introducing the subset of essential reality. We've seen how we can distinguish essential reality from what is non-essential using what is necessary and sufficient in physics models. We've also found that this essential reality, which is a very small set since it's just at a moment, is consistent with what we can and need to manage in living.

Initial conditions & nowflow

As the name *essential reality* suggests, if we try to
take anything away, something essential is missing.
We've used initial conditions to see how the present
moment is enough to cover the whole effect of
all the other moments in physics models. We've
also used initial conditions to see how the present
moment, now, is enough in this way *only if* now
includes the flow of changing, as the word *nowflow*
reminds us. (In contrast, a moment as now-shape,
i.e., including just the position of things in space at
the present, would be insufficient, as initial conditions in physics makes clear.)

Since the essential reality of nowflow is sufficient
for all effects in physics models, even acceleration
cannot add an additional effect once essential reality
is given. Similarly, past events themselves are not
part of essential reality and so are non-essential,
as are future events themselves. Though essential
reality is only at a moment, which is seemingly so
brief, it is enough to cover the whole effect of 4D
spacetime in physics. Deterministic models are used
for clarity in this consideration. The applicability of
the essential reality of nowflow is then generalized
to include indeterministic cases.

Managing nowflow & illusion of separation

In Chapters 2 and 3, we applied the essential reality of nowflow to our living and saw ways that it is consistent with our experience. In the example of standing up quickly, we saw how our movement was more effective when we felt how we are managing nowflow's force flowing form. When trying to manage something non-essential like acceleration, though, we tended to overlook what is crucial for our movement, such as missing the force at our feet and straining ineffectually upward with our head and upper body.

In the example of hurrying to a meeting in Chapter 3, we also saw how trying to manage more than the essential reality of nowflow can make us miss something important for our action. Instead of recognizing that it's the story future about our arrival that's affecting us now, we might have the illusion of separation from nowflow as though we're dealing with the future event itself of our arrival, which hasn't happened yet.

This illusion of separation from nowflow can make us miss crucial aspects of essential reality for our action, such as presently oncoming traffic as we cross a street. When we look more fully into our experience, we see how the story past about our departure, story present about crossing the street, and story future about our arrival are all happening now as information in our mind, which is how they are available to inform our actions now.

Presentism-like essential reality

Since our living is always happening in the present moment, a view that "everything is now" would seem closer to what we experience. The view of presentism is more intuitive in this way since it's the view that all of reality is just at the present moment. Presentism, though, faces some objections as a valid view of reality, such as being incompatible with physics.

The essential reality of nowflow is like presentism in that everything in essential reality is just at a moment. The essential reality of nowflow, though, covers the same whole effect as 4D spacetime and so is sufficient as a view of reality that's compatible with physics, including relativity. As described in Chapter 4, we saw how presentism-like essential reality resolves these objections to presentism, including the question of time.

Time

In presentism, it's difficult to see what would be a valid view of time since reality is only at a moment. If "everything is now," then it would seem that time is not part of reality or that the reality of time must be fully included in the now. By starting with essential reality, which is only at a moment, we've seen how all the effects of time can be covered without a starting assumption about the reality of time. We've also seen how essential reality time is fully part of the essential reality of nowflow.

As we saw in Chapter 5, comparative changing gives all the effects of time, including clocks and their time values. Clocks are comparatively changing things chosen as a standard measure of change, which provides time values for the story past, present, future. Instead of starting with the dimensional time of 4D spacetime, we saw how essential reality starts with comparative changing.

In physics as well as in our living, we may usually construct a view of 4D spacetime so that we think we're managing space and time. Dimensional time, which extends beyond a moment, is not part of essential reality, though, and isn't required for the effects of time. In this way, dimensional time is non-essential, which could be interpreted as there is no dimensional time.

Instead, what is necessary and sufficient for all the effects of time is essential reality with its comparative changing. The comparative changing of a clock in essential reality gives time values as quantitative information for the story past, present, and future about changes. Essential reality time is change information coming from a clock, which is a chosen changing thing. All of essential reality time is part of essential reality, even as time values quantify the changes beyond the essential reality of space and chang*ing* at this moment.

(Essential) Reality as nowflow

In this way, a view of reality as being at one moment can be a valid view. In the view of reality as the essential reality of nowflow, reality includes space and comparative changing at this moment. We've seen how reality in this view includes nowflow's force flowing form. It includes the story past, present, future and time value information. Essential reality also includes essential reality time, which is the change information from the comparative changing in nowflow, specifically from a clock.

The past itself, the future itself, and dimensional time are not part of the essential reality of nowflow so, if we take reality to be the essential reality of nowflow, this can simplify our view of reality. This also supports a simpler view of living, which can make managing our lives easier.

If "everything is now," then there's nothing in reality outside of just this moment, now. In this way, the view of reality as nowflow naturally lessens the illusion of separation from nowflow. It's all just at one moment, now, that includes the flow of changing. By showing how one moment is sufficient, essential reality gives us this option for a simpler view of reality that's both consistent with physics and helpful in living.

References

1. Chabay, Ruth and Bruce Sherwood. 2015. *Matter & Interactions.* John Wiley & Sons.

2. Emery, Nina, Ned Markosian, and Meghan Sullivan, "Time", *The Stanford Encyclopedia of Philosophy* (Winter 2020 Edition), Edward N. Zalta (ed.), URL = <https://plato.stanford.edu/archives/win2020/entries/time/>.

3. Power, Sean Enda. 2021. *Philosophy of Time: A Contemporary Introduction.* Routledge Taylor & Fancis Group.

4. Einstein, Albert. 2015. *Relativity: the Special and General Theory.* Princeton University Press.

5. Carroll, Sean M. 2019. *Spacetime and Geometry: An Introduction to General Relativity.* Cambridge University Press.

6. Westphal, Johnathan and Carl Levenson, eds. 1993. "Aristotle, 'Time' from the Physics" in *Readings in Philosophy: Time.* Hackett Publishing Company.

About the Authors

Wonchull Park is a physicist and tai chi master who brings the rationality of science and the practicality of martial arts into a comprehensive understanding and practice of optimal action. Dr. Park founded Wuwei Tai Chi School (www.WuweiTaiChi.org) while a principal research physicist at Princeton University.

Mackenzie Hawkins received her BA from Princeton University, where she studied the history of science and philosophy, and works as a writer, curriculum designer, and consultant. She especially loves teaching tai chi and meditation in nature.

Reach the authors with comments and questions via email:
nowflow.mackenzie@gmail.com
nowflow.park@gmail.com

www.ingramcontent.com/pod-product-compliance
Lightning Source LLC
Chambersburg PA
CBHW021139020426
42331CB00005B/830